THE
TRUTH
ABOUT OUR WORLD

PALMETTO

PUBLISHING

Charleston, SC

www.PalmettoPublishing.com

Copyright © 2024 by Mary K. Eastman

Paperback ISBN: 979-8-8229-4238-7

THE
TRUTH
ABOUT OUR WORLD

MARY K. EASTMAN

FOR MY SISTER,
CONNIE CANALE,
WHO HAS BEEN WITH ME
EVERY STEP OF THE WAY ON
THIS MIRACULOUS JOURNEY
TO LEARN THE TRUTH.

FOREWORD

A few years ago, my life and the way I looked at things changed. I think COVID-19 changed most of us in some way. It opened my eyes to what was going on in the world. The lies about COVID and the lies about the vaccine made me wake up and start doing some of my own research. I was shocked at how evil our world leaders had become. It was finally discovered that Dr. Anthony Fauci paid Chinese government leaders to make this virus that was "accidentally" unleashed on the world. Once this virus was discovered, the media and the world stopped talking about how it got started. There was no accountability for what they had done. If it had been you or I that made a virus that affected people around the world, we would be in jail, or we would have been executed. For them? Nothing. None of the leaders around the world protested. This told me that either they were all in on the deceptions and COVID lockdowns or they were afraid of the ones that were.

I'm telling you all this because I started seeing the world with new eyes. The scales had come off, and I began to realize that things I had been taught all my life were false. We were being brainwashed by our governments through our schools, our news media, Hollywood, our politicians, and other government agencies. The elites and leaders of our world are all working together.

I must add here that some people are unknowingly being used. For example: teachers have been deceived into believing the earth is round and spinning just like the rest of us were deceived. They, in good faith, were teaching our children what they believed to be true.

What is the end game of the people who control the world? I'm not positive, but I know it's not good. We will look into the possibilities together later in this book.

I've come to the conclusion that the only thing we can rely on totally is God's Word. God is our Creator and he made our world. I disregard anything I hear that contradicts what the Bible says because God's Word has never failed me yet.

I will begin this book by clarifying a few things first, and then I will get to the undeniable proofs that the earth is not spinning and it isn't a ball. In the second part, I will cover Bible verses in regard to our world and the sky above us. In the last part, I will go over scientific evidence and answer some commonly asked questions.

When the fact that our earth is stationary and flat became known to me, it changed my life. My trust and belief in God and his Word has solidified, and I no longer fall for the deceptions of the world. Things we've been told about the moon and stars are not true either. Today, when I look at them, I look with awe and wonder. They are so close to us, and God is on his throne above us and is watching over us. I am truly amazed that he cares for us that much.

I want to share with you the things I have discovered. You will be pleasantly surprised.

CONTENTS

FINDING THE TRUTH IN NATURE AND SCIENCE

Flat or a Round Ball? That is the Question

Most of us were taught all of our lives that we lived on a round ball that was spinning and hurtling through space even though we feel none of this motion. We were kids so we just believed what we were told.

In 2022 I was presented with another possibility: the possibility that the earth was flat and stationary. I couldn't picture in my mind how a flat earth would work or what it would look like, so I laughed at it. Since then, I have learned how this concept works, and I want to describe it to you.

The earth is basically flat. Of course, it has mountains, hills, and valleys, but it is on a flat foundation and it's not tipping, spinning, or moving upward. God tells us that there is a dome, or vault, over the earth. My belief is that the outer edges are in the shape of a square, but this is my theory, and I base it on what I read in the Bible. They won't allow us to go below the 60-degree parallel line, so we can't verify this.

Inside the square is the Antarctic circle. It is a huge wall of ice that goes around our continents and oceans; holding the waters of the oceans in. The waters are not flowing off the edge of the world. In the center of our world is the north pole with Polaris faithfully stationed over it. The sun, moon, and stars are inside the firmament. There is nothing below our earth except its pillars, and there is nothing outside the dome, except heaven and all that is in it.

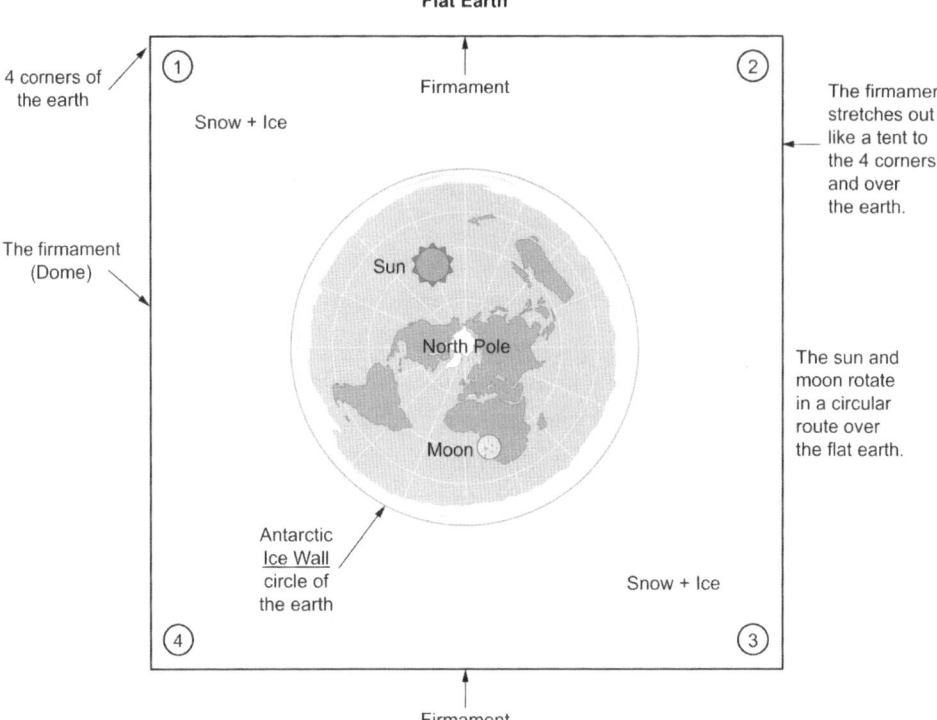

Flat Earth

4 corners of the earth

① Snow + Ice

② The firmament stretches out like a tent to the 4 corners and over the earth.

Firmament

The firmament (Dome)

Sun

North Pole

The sun and moon rotate in a circular route over the flat earth.

Moon

Antarctic Ice Wall circle of the earth

Snow + Ice

④

③

Firmament

More Questions

It's good to question things and explore new possibilities. It's the only way to know what's true and what isn't true. I'm going to start our journey to finding answers by listing some of the questions I had, and I will be addressing some of them in this book. Here they are…

If we went to the moon already, why are we now having trouble crossing the Van Allen Radiation Belt?

If we went to the moon in 1969, why haven't we been able to make it back there?

Why is it that only governments or government agencies can make world maps?

Why can't anyone go to the South Pole? Why can't we go below the 60-degree parallel line?

Why are there no-fly zones?

Why do all flights have to be registered and approved by the government?

Does the moon spin while it's orbiting us?

If there's been so many people and satellites up in space, why don't we have one real video of the earth spinning?

Why doesn't any country own the South Pole/Antarctica?

If all this was not planned by an intelligent creator, why do the constellations and everything remain so predictable year after year? Why don't they change shape, brightness, or become distorted?

Why hasn't Polaris, the North Star, ever moved or changed?

Why is Antarctica so heavily guarded?

Why was the Antarctic shut off by military the same year NASA was formed?

Why do we need military bases to protect uninhabited ice?

If we don't live on a flat plane, why is this place called a planet? Instead of having "plane" as the root word, shouldn't this be called a "globet"?

Why do compasses always point north?

What proves the earth is round or spinning?

Are there really satellites or are they bouncing cell phone signals off the dome?

Why is gravity selective about what it holds to the earth?

Why aren't there any real images of Earth from space?

How do you get time-lapse photos of the stars if the earth is spinning?

Water is always level. Is this why there aren't any *real* photos of the ocean bending?

If you can get motion sickness traveling in a car, why don't you get motion sickness on a globe that's spinning over 1000 mph?

How can a helicopter hover over a stationary object? Wouldn't the earth's spin move it away from them?

If everything is spinning with the earth inside its atmosphere, why do clouds and the wind go in various directions?

If gravity holds oceans to the earth, how can we effortlessly raise our arms and legs?

If the sun is millions of miles away, how can our limited vision see clouds moving behind it?

Wouldn't flight paths become confusing on a spinning ball? Wouldn't it be hard to land on a target that's moving?

If the sun is millions of miles away, why can we see the different angles of its rays?

When they landed on the moon in 1969, who took their picture?

Why did they shoot so many rockets into the sky during and around the 1950s and 1960s? Were they trying to see how high and wide the dome was?

Rainbows are always curved. When there are two rainbows, the colors are inverted. Is it because the rainbow is reflecting off the dome that's a mirror of cast bronze? *Job 37:18*

These are just some of the questions that will get a person thinking. Before getting to the evidence, I have one more item I have to clear up. It's the matter of who "they" are.

Who Are "They"?

When writing this book, I often used the term "they." During the editing process, I was asked who "they" were. That was an excellent question and a hard one to answer, but I'll try.

"They" are the people of this world who are in positions of power and influence, and they know the earth is flat and stationary, yet they present the world to us as being round and spinning. What makes it hard to explain is that "they" sometimes means all of them and sometimes it means just some of them. For example: when talking about the fake South Pole, "they" usually means the leaders of the twelve countries that signed the Antarctic Treaty making it illegal for ordinary citizens like you and I to go below the 60-degree parallel line without their permission. They always have military there to make sure no one goes there on their own. We can't see for ourselves if the world is round or if it has an end. They want us to take their word for it. The only thing they allow is going on one of their approved tours to a peninsula in Antarctica. They let you see some penguins, etc., and then you have to leave.

Sometimes "they" includes world organizations like the WHO (World Health Organization) and the WEF (World Economic Forum). They will never admit it, but they believe the earth is flat. I know this because their logo is a picture of a flat earth map.

Most, but not all, of the world leaders and the powerful and wealthy are in on this deception about our world. They know the truth about our world, but they are encouraging the lie that we live on a spinning ball.

Next is the mainstream news media. The rich and powerful own the majority of the stock in most corporations, especially the media outlets. This gives them control over a good chunk of everything in this world including the information we receive.

Now you're probably wondering why they would be lying about all of this. What does it matter? I get to the "why" in the last section of this book. Here I am only addressing who "they" are.[1]

Now for the fun part: the proof that we're not living on a spinning ball.

---

[1] Wolf Watch: Impossiball https://totaldisclosure.net?impossiball/?doing_wp_cron=1708207810.36 10019683837890625000 Total Disclosure Films 2017, 1:34:27 minutes (start at the 1:08:21 mark)

Airplanes

I'm starting with airplanes because they prove flat earth. Many international pilots have admitted the earth is flat. Most of them will not publicly admit it because they're afraid of losing their job.

This is my best plane argument for flat earth: If a plane is going a little over 575 mph east and the earth is spinning in that same direction around 1000 mph, how will the plane ever reach its destination? It can't. The destination will get so far ahead of the plane that the plane will never get to where it wants to go.

If a plane is flying halfway around the world, the direction you choose to go would make a huge difference in your traveling time and how much fuel you use. If you travel west, the earth's spin will bring the destination your way and you will be flying towards it. If you go east, you will be chasing your destination. Why is it, then, that it doesn't matter which direction they go? Why is it that a plane going from California to New York takes the same amount of time that a plane going from New York to California does? A spinning, round earth doesn't make sense.

Pilots would also have to be adjusting their altitude to follow the downward curve or else they would keep going straight out into space. One person's argument was that the plane must be set for a certain altitude above Earth and automatically adjust itself. If a plane did that, they would constantly be going up alongside hills and mountains and then down again on the other side. Crazy.

Another thought I had was landings and take offs. Wouldn't it be hard to land a plane on something that is spinning over 1000 mph? And wouldn't it make it even harder to land a plane on a runway that goes north and south? You would get lined up for landing the plane, but the runway would keep moving east, away from you. A round, spinning earth would make flying treacherous.

Add to all that the supposed fact that the earth isn't just spinning; it's moving forward and around the sun. Wouldn't it be a nightmare for the person scheduling all these flights around each other? They would have to be calculating the earth's spin and movement forward to coordinate all the plane landings, flights, and take offs.

Also, when the wheels of a plane touch down on something moving that fast, especially if one wheel touches before the other, wouldn't it send the plane into a spin when it touches Earth?

Another thought to consider is that planes have to go east or west. They won't let a plane fly over the North Pole or the South Pole. It's illegal. These are no-fly zones. I find that interesting. Are they hiding something?

Just one more thought on this topic. Why did they call these objects that fly through the air "airplanes"? Shouldn't they be called air globes, or airballs? They named it the way they did because they're flying over a plane (flat surface).

Water

In order to be contained, water needs a container. If water gets spilled, it will always seek the lowest level and it always lays flat.

I have seen many ponds and lakes that have been so still and calm; they were like a mirror reflecting whatever was on the shoreline. It could be a city, mountain, or a bunch of trees. If there is no wind or boats or anything else stirring the water up, it always lies flat and it doesn't move.

When I look at a globe now or a picture of the round earth with the oceans curving, it looks totally ridiculous to me. How did I ever think that was real? Yet, that's what I've believed all my life because the "authorities" on the subject said so.

If the earth is spinning over 1000 mph, why aren't we and the oceans and everything else flying off? Oh, I know. You're going to say, "Gravity!" I guess I need to talk about that next.

Gravity or Density?

The definition of gravity is that it's a force attracting a body to the center of the earth or toward any other physical body having mass. It's never been proven that gravity exists. It's just a theory. The only proof they have is that a man drops an apple and it falls to the ground. This proves nothing. Gravity is something they made up to supposedly explain why everything is held tightly to an earth that's moving crazy fast, yet nothing moves because of it or flies off.

This "gravity" is so strong that it keeps oceans from sloshing around and it holds everything to the ground. It holds ants, people, buildings, and tanks fast to the ground, yet birds and butterflies can defy this profoundly powerful source. They can walk on the earth or defy gravity and fly around in the sky. They're pretty amazing.

The law of density explains why things will fall to the ground and why things less dense than air will rise up. Density is the measurement of how tightly a material is packed together. Something denser than air will naturally sink. No gravity is needed, and gravity's never been proven to exist, yet it is spoken of as if it's a fact.

Time-lapse Photos

I love looking at pictures of the sky. Lately, I have come across time-lapse photos of the stars. They are easy to find online.

The stars are all going in a circle over our flat earth with the North Star (Polaris) stationary and in the middle! They wouldn't look like that if we were on a ball spinning wildly through space. The stars would be random.

The other thing I found interesting is that Polaris, the North Star, is always at the north (center of the flat earth). Why doesn't it ever move? With a spinning world that's hurtling through space, it might have a hard time keeping up!

The North Pole

What's at the North Pole? Most people would say snow and ice, nothing more. Some might be funny and say Santa's Workshop. If you've ever watched expeditions to the North Pole, it looks miserably cold. No one would ever want to go there.

I do have some questions about that though. Compasses always point north. I'm not a scientist, but I'm thinking that there has to be some magnetic force at the North Pole that would draw the needle on a compass that way. Even in the southern hemisphere, compasses point north. They won't let us go up to the North Pole, which makes me wonder why. Is there something they don't want us to see? If so, then what?

Also, we have the aurora borealis (northern lights) to think about. I'm sure scientists came up with a reasonable explanation for this, but there's something going on at the North Pole and no one seems to be able to go there to check it out. What is really causing the northern lights? I would like to know.

I searched on the internet and they said yes, people can go independently to the North Pole. However, I've watched many videos where people were saying no one is allowed up there. If you get close, the military shoos you away. Supposedly, in Russia there is a place you can go, and they will take you to the North Pole, but how do we know we're actually at the North Pole and not just some snowy, ice sheet that they're calling the North Pole?

Why am I so skeptical? I guess it's because we've been lied to so many times that I find it hard to believe anything the authorities say who control the information we receive anymore. My gullible days of believing the given narrative are over. I want proof.

There are ancient maps that show a flat earth and the North Pole is at the center. There is a huge black rock (or mountain) in the center.[2]

---

[2] AE Realm ForHIsNameSake, https://bitchute.com/video/o7NcjvgvNwnC/ Ancient Maps of North Pole. Published on February 27th, 2023. 6:15 minutes

Around it is swirling waters because the four continents around it are divided by four rivers flowing into the center. Many people believe that the water is getting sucked into a hole which goes down into the earth. I'm guessing the water flows through underground rivers and goes back out through the world. If these old maps are accurate and the accounts of a whirlpool that would suck ships down into it are real, it would make sense that they would want to keep us away from there for our safety; but why don't they just tell us that? Why so much secrecy about the North Pole?

Do these continents show up on maps now? Nope. Maps show nothing up there now but floating ice blocks. (World maps are made by the government). The continents just disappeared. I suppose it's possible they're under water now or maybe they don't want us going to those continents and seeing what's up there. We don't know.

Some ancient drawings of our world show a flat earth with a dome over it. In the center of the world, some of these drawings show a huge tree in the center that goes all the way up to heaven at the top of the dome. This reminds me of a story in the Bible in Daniel, chapter four. In the dream, there was a huge tree in the middle of the land, and its top touched the sky and was visible to the ends of the earth. (Notice that the world has ends. It is not a ball.) The tree gets cut down and all that remains is the stump and its roots, bound with iron and bronze. The interpretation of this dream is that the tree is Nebuchadnezzar, and he loses his power until he admits the Most High is sovereign over all the land. This story is about Nebuchadnezzar. Is it possible that it has a hidden meaning? Many stories in the Bible have a second or hidden meaning. One theory that is discussed says that there is a black rock at the center of the world. Could it be the stump bound with iron and bronze? Was it originally a huge tree that was cut down? Does it have some special magnetic power that pulls the needle on compasses toward it? We don't know. This is one of endless theories.

Years ago, my adult son moved back in with me for a while so he could put money into starting a business. One day my grandkids were coming over, and he told me not to let them go in his room. He likes to hunt and fish so there were things they could get hurt on. Also, his room was a little messy. When the kids came over, I told them they could go in any room in the house, but not their uncle's room. What do you think telling them that did? It made them want to see why they couldn't go in there. They started imagining different reasons. I saw this wasn't going to die a natural death, so I opened the door and let them look in.

"It's a little messy, isn't it?" I asked them. They giggled and agreed, and they never asked to go in there again.

The reason I'm telling you this is that all people are naturally curious. Why all the secrecy about the North Pole and the fake South Pole? Just let people go to these places and the truth will be known, and it will be put to rest.

NASA—1958

What does NASA stand for? National Aeronautics and Space Administration. What does *nasa* mean in Hebrew? It means to delude, to beguile, to deceive or seduce. Basically, *nasa* means lies. You might say, "That's just a coincidence. They didn't do that on purpose." Well, Hebrew is the first language of the Jewish people who were known as Hebrews or Israelites back in the days of Moses. A lot of the people at the top today are Jewish. I don't believe for a second that their choice of "NASA" was an accident. They knew what *nasa* meant. The elites and world leaders enjoy lying to us and then putting it right in front of our faces. They're laughing at us.

The Moon Landing

We supposedly landed on the moon. How do you land on a light? How did they get inside the firmament where the moon is? (The Bible says the moon is a light and it's in the firmament).

I was taught all my life that the moon is not a light. It reflects the sun's light. This year it became clear to me this was another lie. If it's reflecting the sun's light, why does it shine brightest when it's in the complete dark with no other light around it and the sun cannot be seen anywhere?

I've seen the moon in the daytime, and sometimes you can barely see it. It blends in with the sky some days, and other days it is more obvious, but it's not shining as bright as it does at night in the dark.

Research the 1969 moon landing on a website that's not censoring content. You will see all kinds of things that don't line up. You will see, if you look carefully, hidden cables helping the astronauts "float." During the filming of the astronauts walking on the moon, you will see their shadows going in different directions like there is more than one light source shining on the moon (such as artificial lights). Also, you will see them taking off and leaving the moon, but who is taking the picture? You will also see a video of President Richard Nixon talking to the astronauts while they were on the moon. He was using a land line![3]

It was 1969. They didn't have technology to talk to people on the moon. Cell phones didn't become common place until the 1990s. Today, we still have dead spots where we can't use our cell phones, yet, transmission to the moon was no problem!

Sometimes I want to kick myself for not thinking about these things or even questioning them. I now look at the moon and common sense tells me it gives off its own light.

---

[3] Trump to Save America Natasha TX, https://www.bitchute.com/video/krPQrTlpHLLk/ Fake Moon Landing Phone call w/Nixon and Men on the Moon, January 22, 2024, 2:09 minutes.

A Return Visit to the Moon?

If astronauts went to the moon in 1969, why haven't they been back? NASA claims they destroyed the technology to get there.[4] Seriously? Something that important and they just destroyed it. Sounds fishy to me.

There was another attempt to go to the moon, and the spacecraft blew up along with the crew that was aboard. We assume the crew was there, but there are some people who have doubts about that. I won't bother getting into that, but it's a convenient reason to not try to go again. After several lives were lost, who would suggest they should try it again? They would look like an inhuman monster.

When talking about going to the moon and other places, they always mention the problem of the Van Allen Radiation Belt and how they can't go through it because of the harmful radiation. So, how did the first astronauts make it through without becoming contaminated? Hmmm. Is this fake Van Allen Radiation Belt they talk about actually the firmament mentioned in the Bible? To admit that, NASA would have to say that what the Bible says about God and our world is true. That's the last thing they will ever do.

Sunlight and Moonlight

Sunlight and moonlight are different from each other. The sun's light is hot while the moon's light is cool. If you are outside at night, it will be cooler in the moonlight and warmer in the shade. This doesn't make sense to me, but it is what it is. I'm okay with just saying I don't know why.

Here is a Bible verse for you that I find interesting.

---

[4] Truth Conspiracy Reality, https://www.bitchute.com/video/QmOP1QOpbdIE/ "We don't have the technology to do that anymore; we used to, but we destroyed it." 11-15-2020, 0:14 minutes.

*Isaiah 13:10 says, "...the sun shall be darkened in "**his**" going forth, and the moon shall not cause **her** light to shine." (King James Version, emphasis added)*

Isn't this interesting that when talking about the sun, the Bible says his going forth. When talking about the moon, it says *her* light to shine. The light from the sun and the light from the moon are vastly different from each other. God's inspired word makes the sun masculine and the moon feminine.

Years ago, when I started working in a factory, we had three shifts. It always amazed me how differently the three shifts acted. First and second shifts were more lively, loud and vocal, much more than third shift.

Back then I liked working overtime, so I would work my second shift and sometimes volunteer to work a few hours on first or third shift. Third shift was more subdued than the others. Yes, they would talk to each other, but they were less rambunctious. I thought it was because they were tired, but that's not accurate. Many of the third shifters had worked on third for years. They were used to it. They had gotten as much sleep as the other shifts.

Also, I noticed a change in myself and others when they switched from one shift to another. I know I enjoyed the liveliness of first shift when I was on first shift, but when I was on a shift that worked into the middle of the night, I didn't desire to talk as much and wasn't as playful in the early morning hours, even if I had plenty of sleep and wasn't tired. Why is that?

My theory is that evening feels different from the day. God speaks often in the Bible about how we, and even the earth, need rest. There are several different kinds of rest. Even God "rested" on the seventh day of creation. (It wasn't that he was tired. He wanted to take time to enjoy all of his work). One example of another kind of rest is that the soil can become depleted of minerals if it's overworked and never gets any "rest." God wants us and even the soil to have rest as the next verse shows.

Leviticus 26:35 "All the time that it lies desolate, the land will have the rest it did not have during the sabbaths you lived in it."

Looking back, I remember the few times I went for night swims in my life. The air and breeze felt different. Sounds seemed to travel through the air and across the water differently at night. Everything seemed so peaceful. My personal belief is that God purposely made the day and night different from each other in several ways so that we could unwind, rest, and sleep in the dark, and be rejuvenated when we woke up.

It fascinates me that the sun's light can be so powerful and seems to energize people, while the moon's light will have a relaxing effect.

I'll add a few more verses here with pronouns for the moon.

*Ezekiel 32:7: "And when I shall put thee out, I will cover the heaven, and make the stars thereof dark; I will cover the sun with a cloud, and the moon shall not give **her** light." (KJV, emphasis added)*

*Matthew 24:29: "Immediately after the tribulation of those days shall the sun be darkened, and the moon shall not give **her** light, and the stars shall fall from heaven, and the powers of the heavens shall be shaken." (KJV, emphasis added)*

We tend to think of people and animals as being male and female, but not many other things. God thinks differently than we do. Many plants and some trees are male and female. Some trees have flowers that have both male and female reproductive parts. God often refers to the church as his bride, and Jesus is the bridegroom. God has a broader use and meaning for the words *male* and *female* than we usually do, so I don't find it too unusual that he refers to the sun as "he" and the moon as "she."

Speaking of God, in the next section we will see more of what our Creator has to tell us.

STRAIGHT FROM THE BIBLE

The Firmament

I'm going to start at the beginning of the Bible in Genesis. Some translations use the term *vault*. Others will use *firmament, dome*, canvas, or even *tent* to describe the covering over our earth.

Genesis 1:6–8: "And God said, 'Let there be a vault between the waters to separate water from water.' So God made the vault and separated the water under the vault from the water above it. And it was so. God called the vault 'sky.' And there was evening, and there was morning—the second day."

We live beneath a vault in a closed system. Some will compare this to a terrarium or those snow globes you see, where you have a scene (and it's flat) and there is a clear covering (or dome) over it. Have you ever watched on TV rockets being shot into space? At first, they always look like they're going straight up, but if you keep watching, you'll see that it starts to curve. If it keeps going straight up it will hit the "glass" ceiling (vault). This next verse shows the texture of the firmament. It's hard and solid.

Job 37:18: "Can you join him in spreading out the skies, hard as a mirror of cast bronze?"

NASA can send all the rockets they want up, but they're never getting through that barrier. As you will see with the next verse, the sun is inside this tent or firmament.

Psalm 19:4-6: *"Yet their voice goes out into all the earth, their words to the ends of the world. In the heavens God has pitched a tent for the sun. It is like a bridegroom coming out of his chamber, like a champion rejoicing to run his course."*

The sun is inside the tent (firmament). It is the domed "ceiling" over our world.

Lights in the Vault

The Bible talks about lights inside the vault. This would include the sun, moon, and stars.

Genesis 1:14: *"And God said, 'Let there be lights in the vault of the sky to separate the day from the night, and let them serve as signs to mark sacred times, and days and years.'*

The lights are in the vault. No one will ever be able to get to them. No plane will ever accidentally fly or fall into a star. They are locked up in the vault and circling above the flat earth. Look up some time-lapse photos of the stars and you will see the circular pattern of the stars above us. There is such precision and predictability in the pattern of the stars. If the earth were spinning rapidly and moving forward around the sun and hurtling through space, Polaris would not be able to remain positioned exactly over the north pole with the stars consistently moving around Polaris.

It also says the lights serve as signs. I don't know much about this yet, but I hope to learn more about what they all mean in the future.

By the Light of the Moon

The next verses show that the moon is a light, but before you read those, I want to explain how the sun and moon rotate above an earth that is flat. The earth is not moving. It's the sun and moon that are taking a circular route above the flat earth. This route isn't always the same. If you've ever noticed, some times of the year the sun is higher in the sky than at other times. The sun and moon sometimes are higher in the sky than other times and sometimes the circle route they take is wider and may be further in the southern hemisphere than the northern hemisphere. Other times of the year it will be the opposite. These slight variances give us the different seasons. Now we'll get back to our moon that is a light.

Genesis 1:16: *"God made two great lights—the greater light to govern the day and the lesser light to govern the night. He also made the stars."*

Isaiah 60:19: "The sun will no more be your light by day, nor will the brightness of the moon shine on you, for the LORD will be your everlasting light, and your God will be your glory."

Jeremiah 31:35: "This is what the LORD says, he who appoints the sun to shine by day, who decrees the moon and stars to shine by night, who stirs up the sea so that its waves roar—the LORD Almighty is his name."

Whenever I read the Bible, I would just zoom right over these verses, but look how it's telling us the moon is a light. I was always told that the moon was not a light itself; it reflected the light of the sun. Is that true? No, it's not. When the moon is shining brightest is at night. There is no sun or sunlight anywhere in sight because the earth is blocking the sun's light. (This is how it would be if the globe model is true).

The moon is giving off its own light. Next time you see a full moon at night, take a good look at it and you will have to admit it is a light.

Also, during the day when the moon is near the sun, it's often a bluish color and blends in with the sky. If it reflects the sun's light, wouldn't it be brighter when it's close to the sun?

The next question you might have would be, "Then how did the astronauts land on it?" All the pictures of the moon landing in 1969 don't show a glowing moon or that it's a light. Today, many are doubting that we ever really went to the moon. This is another subject I looked into, and I was shocked at what I found out. I will cover this more in the last part of this book.

The Tower of Babel

I'll give you a brief recap of the Tower of Babel as told in the Bible; in case you're not familiar with it.

God had told the people of those days to spread out, multiply and fill the earth. God never intended for everyone to stay right where they were at. He wanted all the continents and islands occupied with people.

The people of those days kind of liked it where they were at. They rebelliously defied God and stayed where they were at. In fact, they decided they didn't need God to get to heaven; they would build a pyramid and walk right on in to heaven all by themselves!

The next verse shows what their thoughts were about this.

Genesis 11:4: "Then they said, 'Come, let us build ourselves a city, with a tower that reaches to the heavens, so that we may make a name for ourselves; otherwise we will be scattered over the face of the whole earth.'

The people of those days thought that they could build a tower high enough to reach heaven. Isn't that ridiculous?

Genesis 11: 6–7: "The LORD said, 'If as one people speaking the same language they have begun to do this, then nothing they plan to do will be impossible for them. Come, let us go down and confuse their language so they will not understand each other.'"

Wait a minute! It sounds like God believed they could achieve this goal, so he set out to make it impossible for them. Wow. God is saying that heaven is that close to us!

The Heavens and the Earth

Since we're talking about heaven, I want to get a quick explanation of heaven in. I'll first have you look at this verse.

Genesis 1:1: "In the beginning God created the heavens and the earth."

Notice that Earth is singular; there is only one Earth, but heavens is plural. The heavens aren't usually numbered in the Bible, except I did find that a third heaven was mentioned once and also a first heaven. So, there are at least three heavens.

*2 Corinthians 12:2–4: "I know a man in Christ who fourteen years ago was caught up to the **third** heaven. Whether it was in the body or out of the body I do not know—God knows. And I know that this man—whether in the body or apart from the body I do not know, but God knows—was* *caught up to paradise and heard inexpressible things, things that no one is permitted to tell."*

Because of this verse and others, I'm convinced that God's throne (paradise) is in the third heaven. The next verses are about the sky.

Revelation 21:1: "Then I saw a new heaven and a new earth, for the first heaven and the first earth had passed away, and there was no longer any sea."

Isaiah 55:10: "As the rain and snow come down from heaven, and do not return to it without watering the earth and making it bud and flourish, so that it yields seed for the sower and bread for the eater..."

James 5:18: "Again he prayed, and the **heavens** *gave rain, and the earth produced its crops."*

The next verses are about the second heaven, the firmament (or vault).

Genesis 1:16–17: "God made two great lights—the greater light to govern the day and the lesser light to govern the night. He also made the stars. God set them in the vault of the sky to give light on the earth."

Psalm 8:3: "When I consider your **heavens**, *the work of your fingers, the moon and the stars, which you have set in place." (Emphasis added)*

Some Bible translations use the term vault and some use the term firmament; so, this is my conclusion about the heavens. The first heaven is the sky where the birds fly, etc. The second heaven is the firmament, and the sun, moon and stars are in the firmament. The third heaven is where we long to go. This is where God's throne is, and the angels dwell there also.

When you come across the word *heaven*, notice if it's in the plural or if it's singular. That will help you know if it's referring to one "heaven" or all three. (If there's more than three heavens, the Bible doesn't mention them).

Jacob's Ladder

Another story in the Bible is about Jacob and what happened while he was on a journey to another country. He had a dream and this is the vision that God allowed him to see.

Genesis 28:12: "He had a dream in which he saw a stairway resting on the earth, with its top reaching to heaven, and the angels of God were ascending and descending on it."

Genesis 28:17: "He was afraid and said, "How awesome is this place! This is none other than the house of God; this is the gate of heaven.""

This is the story of Jacob's dream. In his dream he sees a ladder from Earth to heaven. Heaven is closer to us than we think!

The Sun and Moon Stood Still

In the Bible, Joshua was leading a battle, and he asked God to help him by making the sun and moon stand still so that they could avenge themselves on their enemies. God granted this request.

Joshua 10:12–13: "On the day the LORD gave the Amorites over to Israel, Joshua said to the LORD in the presence of Israel: 'Sun, stand still over Gibeon, and you, moon, over the Valley of Aijalon.' So the sun stood still, and the moon stopped, till the nation avenged itself on its enemies, as it is written in the Book of Jashar. The sun stopped in the middle of the sky and delayed going down about a full day."

The next two verses show the movement of the sun.

Ecclesiastes 1:5–6: "The sun rises and the sun sets, and hurries back to where it rises. The wind blows to the south and turns to the north; round and round it goes, ever returning on its course."

Isaiah 59:19: "From the west, people will fear the name of the LORD, and from the rising of the sun, they will revere his glory. For he will come like a pent-up flood that the breath of the LORD drives along."

The sun and the moon are moving, which is a common belief of most people. I shouldn't get much argument here; both the heliocentric and geocentric models say the sun and moon are moving. The only thing to notice here is that Joshua doesn't ask to have the earth stop. It's stationary.

Earth's Foundation

The earth has a foundation. This only works on a flat earth; not a ball earth.

1 Samuel 2:8: "...For the foundations of the earth are the LORD'S; on them he has set the world."

Zechariah 12:1: "The LORD, who stretches out the heavens, who lays the foundation of the earth, and who forms the human spirit within a person declares."

Isaiah 42:5: "This is what God the LORD says—the Creator of the heavens, who stretches them out, who spreads out the earth with all that springs from it, who gives breath to its people, and life to those who walk on it..."

The earth is spread out. It is on a foundation. A spinning ball cannot be set on a foundation. When I first heard that the earth was flat, I laughed because I couldn't picture it. No one explained to me how the earth is set up, so it made no sense. I was trying to picture a flat earth with the North Pole at the top and the South Pole at the bottom. It wasn't working in my mind because that's not how it is.

This is how God set the earth up. Picture the North Pole in the center of a flat earth. The continents and oceans surround it. The outer wall is a huge ice wall that goes all the way around holding the water in. (Water needs a container.)

There is no South Pole. The fake South Pole (which is really the ice wall around the whole outer edge) is constantly guarded by the military. No country or person owns the South Pole because it doesn't exist. There was an Antarctic Treaty System signed in 1959 where twelve countries agreed to not let people go to the "South Pole." If you try to fly over it or take a boat to land on it, you will be stopped. No one is allowed, without their permission, to go below the 60-degree parallel line.

While I used to laugh when I thought of a flat earth, I now chuckle when I see a picture of a round earth with curving oceans. We've all seen ponds and lakes when they were so still that they looked like a mirror. Water will always flow downward, seeking the lowest level, until it's contained. The oceans are flat too. When I look at a globe or a picture of a round earth, I now see how ridiculous it is. Water doesn't curve like that. If you pour water in a glass, it will always be level.

The Earth Can't Be Moved

Is the earth moving? Let's see what our Creator says about that.

1 Chronicles 16:30: "Tremble before him, all the earth! The world is firmly established; it cannot be moved."

Psalm 96:10: "Say among the nations, 'The LORD reigns.' The world is firmly established, it cannot be moved; he will judge the peoples with equity."

Psalm 104:5: "He set the earth on its foundations; it can never be moved."

The earth is not moving, yet you will be told that the earth is spinning over 1000 mph and the moon is circling around the earth, and both of them are going around a sun that's moving. Wow! It amazes me that I believed all that for sixty years! When you're a kid, you will believe anything adults tell you, even that a man named Santa Claus delivers toys to kids around the whole world in one night. Amazing.

If the world is spinning like that, why is it we don't go flying off? They will tell you there's a magical force called gravity, but gravity doesn't exist. It's the law of density that determines what falls to earth and what floats up; things more dense will sink and things less dense will rise up. Gravity has never been proven to exist. It's just a theory.

Earth's Pillars

What's under the earth? Good question. Let's find out.

Job 9:6: "He shakes the earth from its place and makes its pillars tremble."

Job 26:7: "He spreads out the northern skies over empty space; he suspends the earth over nothing."

Psalm 75:3: "When the earth and all its people quake, it is I who hold its pillars firm."

There is nothing under the earth except its pillars, and God uses the pillars to hold the earth when it shakes. God has a "handle" on things!

Job 26:11: "The pillars of the heavens quake, aghast at his rebuke."

I'm just adding this because I found it interesting that the heavens have pillars too.

The Face of the Moon

The moon has a face, so there is a man on the moon! (I'm kidding, I'm kidding.)

Job 26:9: "He covers the face of the full moon, spreading his clouds over it."

Seriously, the Bible uses the term "face of the earth" often and we know the earth has a surface but nothing under it. If the moon has a face, does that mean it's not a ball and there's only one side to it? You always hear the term "the dark side of the moon." Do they say that because they know it's not spinning, and it has only the one side (or face) that we can see and the back side?

The Horizon

The horizon is the boundary between light and darkness, day and night.

Job 26:10: "*He marks out the horizon on the face of the waters for a boundary between light and darkness.*"

Proverbs 8:27: "*I was there when he set the heavens in place, when he marked out the horizon on the face of the deep.*"

God Sees the Ends of the Earth

God can see all of mankind and he views the ends of the earth.

Job 28:24: "*…For he views the ends of the earth and sees everything under the heavens.*"

Psalm 33:13: "*From heaven the LORD looks down and sees all mankind.*"

Matthew 4:8: "*Again, the devil took him to a very high mountain and showed him all the kingdoms of the world and their splendor.*"

Revelation 1:7: "*Look, he is coming with the clouds, and 'every eye will see him, even those who pierced him'; and all peoples on earth 'will mourn because of him.' So shall it be! Amen.*"

Before I knew the earth was flat, I believed people who said, "Well, the people on the other side of the earth will see him return on television." That sounds so ridiculous now. Every eye will be able to see him come because the earth is flat. We're not going to be sitting on our couch saying, "Hey Hon, come here. Jesus is on television and he's here."

Earth's Edges

A round, ball earth doesn't have edges. The earth is flat.

Job 38:12–13: "*Have you ever given orders to the morning, or shown the dawn its place, that it might take the earth by the edges and shake the wicked out of it?*"

Boundaries for the Seas

I've seen so many ridiculous images of a flat earth with the waters of the oceans rolling off its sides and boats tumbling perilously off the edge. I would laugh at first; but it's getting old.

Job 38:8–11: "Who shut up the sea behind doors when it burst forth from the womb, when I made the clouds its garment and wrapped it in thick darkness, when I fixed limits for it and set its doors and bars in place, when I said, 'This far you may come and no farther; here is where your proud waves halt'?"

Psalm 74:17: "It was you who set all the boundaries of the earth; you made both summer and winter."

Psalm 104:9 "You set a boundary they cannot cross; never again will they cover the earth."

God creates boundaries for water. The shorelines hold the waters for ponds, lakes, rivers, and oceans in place.

What keeps the water from falling off the ends of the earth? It's the Antarctic ice wall that goes all the way around the oceans and continents. Have you seen images of the Antarctic ice wall? It's huge! It's 200 feet high in some places!

I'm guessing that there's just snow and ice beyond those ice walls until you come to the "glass" dome. The earthly powers that be won't let us go below the 60-degree parallel line, so we just have to guess on that one. (They won't let us go to the North Pole either!) What are they hiding?

The Heavens are Stretched Out Like a Tent

Well, what do you know! God doesn't live on a spinning ball either. The heavens are stretched out.

Psalm 104:2: "The LORD wraps himself in light as with a garment; he stretches out the heavens like a tent..."

Isaiah 40:22: "He sits enthroned above the circle of the earth, and its people are like grasshoppers. He stretches out the heavens like a canopy, and spreads them out like a tent to live in."

This is the only verse I found that referred to the earth as a circle. It never refers to the earth as a ball. Circles can be flat like a plate. The ice wall is a circle holding the oceans in. I'll explain my idea of the shape of the earth next when I share verses about the four corners of the earth.

The Four Quarters of the Earth

The Bible sometimes says "four quarters" and sometimes it says "four corners."

Isaiah 11:12: "He will raise a banner for the nations and gather the exiles of Israel; he will assemble the scattered people of Judah from the four quarters of the earth."

Jeremiah 49:36: "I will bring against Elam the four winds from the four quarters of heaven; I will scatter them to the four winds, and there will not be a nation where Elam's exiles do not go."

Ezekiel 7:2: "Son of man, this is what the Sovereign LORD says to the land of Israel: The end! The end has come upon the four corners of the land!"

Ezekiel 37:9: "Then he said to me, 'Prophesy to the breath; prophesy, son of man, and say to it, 'This is what the Sovereign LORD says: Come, breath, from the four winds and breathe into these slain, that they may live.'""

Revelation 7:1: "After this I saw four angels standing at the four corners of the earth, holding back the four winds of the earth to prevent any wind from blowing on the land or on the sea or on any tree."

We know the arctic wall is a circle that goes around the outside of all the continents and oceans and it's holding the waters in.

However, it doesn't end there. It's not like you climb up the ice wall and you're knocking on the "glass" dome. There's more beyond that. I've seen photos on the internet showing that the ice and snow continues on. Again, they won't let us go out there, but I'm thinking the land goes out into the shape of a square and then you hit the dome. I've never heard anyone else say that, but going from God's Word, that's what I'm thinking.

The Bible uses words like four corners, canopy, and tent. To me, this implies that the very outer edge could be a square. Inside the square is the Arctic Circle. Inside that are the oceans and continents where people live. In the very middle is the North Pole.

Many people think they're keeping us from going out there because they don't want us to know that we live in a closed system and can't get out. If we knew that, it would be proof that the Bible is right and there is a God. They don't want us believing in God.

The Waters from Heaven

God waters our world with water from his upper chambers (and from the clouds, I'm assuming).

Psalm 104:13: "He waters the mountains from his upper chambers; the land is satisfied by the fruit of his work."

Jeremiah 10:13: "When he thunders, the waters in the heavens roar; he makes clouds rise from the ends of the earth. He sends lightning with the rain and brings out the wind from his storehouses."

Sun, Moon, and Stars. Planets?

The sun, moon, and stars are mentioned in the Bible, but are planets mentioned?

Isaiah 45:12: "It is I who made the earth and created mankind on it. My own hands stretched out the heavens; I marshaled their starry hosts."

Jeremiah 8:2: "They will be exposed to the sun and the moon and all the stars of the heavens, which they have loved and served and which they have followed and consulted and worshipped. They will not be gathered up or buried, but will be like dung lying on the ground."

1 Corinthians 15:40–41: "There are also heavenly bodies and there are earthly bodies; but the splendor of the heavenly bodies is one kind, and the splendor of the earthly bodies is another. The sun has one kind of splendor, the moon another and the stars another; and star differs from star in splendor."

I don't know what to think about planets. I was taught all my life that there were nine planets. They're not mentioned even once in the Bible. I'm not sure what the Bible means by starry hosts or heavenly bodies. I'm guessing they're the stars? I remember seeing a news video on the computer once, and they were telling us that the two bright objects in the sky were planets and they were near each other and that was a rare thing. To me, they just looked like tiny stars shining. The pictures I've seen of the planets are just that: pictures. (Not photos.) They don't look real to me.

What I'm getting at is I'm not sure if there are planets. They may be nothing more than wandering stars. Like I said before; I trust God more than I trust our government agencies.

Thrones and Footstools

Kings should have a throne and footstool, so it is reasonable that God should have one.

Isaiah 66:1: "This is what the LORD says: 'Heaven is my throne, and the earth is my footstool. Where is the house you will build for me? Where will my resting place be?'" (referenced in Acts 7:49)

Lamentations 2:1: "How the LORD has covered Daughter Zion with the cloud of his anger! He has hurled down the splendor of Israel from heaven to earth; he has not remembered his footstool in the day of his anger."

Isaiah 66:1 This is what the LORD says: "Heaven is my throne, and the earth is my footstool. Where is the house you will build for me? Where will my resting place be?"

I like Isaiah 66:1. I picture God sitting in heaven on his throne, and he's casually resting his feet on the top of the dome. Maybe he's even laughing with the angels about some hilarious things we're doing. Or maybe he's moved by some big sacrifice someone did to help someone else. I don't know. I guess I just like knowing that he's just beyond the clouds and he's watching us.

Bette Midler said in a song that he's watching us from a distance. While everyone loved that song, I always felt like she was mocking God and was saying he didn't know what was really going on down here. Nothing could be farther from the truth!

When I try to picture these verses with a ball earth that's spinning, the images in my head become hilarious ones. Would God want to rest his feet on a ball that's spinning? I don't think so.

Foundations of the Earth

Several times the Bible mentions the earth having foundations. Foundations are flat. How would you have a foundation on a ball?

I found twelve verses on the foundations of the earth. I'm guessing there are more, but these are ones I will reference here:

1 Samuel 2:8: "He raises the poor from the dust and lifts the needy from the ash heap; he seats them with princes and has them inherit a throne of honor. 'For the foundations of the earth are the LORD's; on them he has set the world.'"

Job 38:4–6: "Where were you when I laid the earth's foundation? Tell me, if you understand. Who marked off its dimensions? Surely you know! Who stretched a measuring line across it? On what were its footings set, or who laid its cornerstone…?"

THE TRUTH ABOUT OUR WORLD

Proverbs 8:29: "When he gave the sea its boundary so the waters would not overstep his command, and when he marked out the foundations of the earth."

Below are the other verses in case you want to look them up:

2 Samuel 22:16/Psalm 18:15/Psalm 82:5/Isaiah 24:18/Isaiah 48:13/ Isaiah 51:13, 16/Jeremiah 31:37/Micah 6:2/Zechariah 12:1

Face of the Earth

God uses the term "face of the earth." If we lived on a ball, wouldn't he say "the head (or ball) of the earth?" A face is only on one side of a person, so I think the meaning here is clear.

There are fourteen "face of the earth" verses that I found. There may be more.

Amos 5:8 He who made Pleiades and Orion, who turns midnight into dawn and darkens day into night, who calls for the waters of the sea and pours them out over the face of the land – the LORD is his name.

Luke 21:35 For it will come on all those who live on the face of the whole earth.

Genesis 7:23/Genesis 11:4,9/Exodus 32:12/Deuteronomy 14:2/Job 37:12/Isaiah 23:17/

Isaiah 24:1/Jeremiah 25:26/Jeremiah 28:16/Ezekiel 38:20/Amos 9: 6,8/Zephaniah 1:2

Ends of the Earth

I found thirty-eight "ends of the earth" verses. This phrase is only suitable for a flat earth because it would have ends. A ball wouldn't have even one end. I won't list all thirty-eight of them here, but I'll list a few of my favorites.

Job 37:3: "He unleashes his lightning beneath the whole heaven and sends it to the ends of the earth."

Psalm 19:4: "Yet their voice goes out into all the earth, their words to the ends of the world. In the heavens God has pitched a tent for the sun."

Psalm 98:3: "He has remembered his love and his faithfulness to Israel; all the ends of the earth have seen the salvation of our God."

Psalm 135:7: "He makes clouds rise from the ends of the earth; he sends lightning with the rain and brings out the wind from his storehouses."

Isaiah 45:22: "Turn to me and be saved, all you ends of the earth; for I am God, and there is no other."

Jeremiah 10:13: "When he thunders, the waters in the heavens roar; he makes clouds rise from the ends of the earth. He sends lightning with the rain and brings out the wind from his storehouses."

If you're like me, the Bible verses are enough to convince me of how our world operates. However, there will be many who will need to see it in the science and I understand that. That's why I wrote the first and third section—to focus on different topics and facts.

Note: I used the NIV translation for my Bible quotes, except where noted.

GLOBAL DECEPTION AND LIES

Deceptive Photography

I have seen thousands of pictures in my lifetime of the horizon, and it's flat every time...almost. Occasionally, I'll see a picture where the horizon is curved. I'm told that a fish-eye lens will bend the horizon like that. Are those pictures no more than trickery?

Sometimes a picture taken from a window on an airplane might kind of look curved. Is it the design of the window that does that? High altitude pictures from a hot-air balloon don't show curvature.

Next, we have to mention NASA. Their official pictures show a perfectly round ball, but if you compare their pictures with ones from other years, the sizes of the continents are different. In one picture, North America looks quite large and in another it looks much smaller. Also, North and South America are mainly what you see. That means every other continent in our world has to be on the other side!

These are not real photos. They've been photoshopped. NASA even admits that they're CGI.[5]

---

[5] Truth Conspiracy Reality, https://www.bitchute.com/video/Q7asRmHLDC12/ NASA's Robert Simmon Admits Ball Earth Images are photoshopped! The Earth is Flat 11/15/2020, 1:53 minutes

Has the Globe Been Proven?

No, the earth being a globe hasn't been proven. There are only a few "proofs" they have. One proof is NASA's photos, but we know now that they've been photoshopped. With all the satellites and trips to space they have made, why don't they even have one video to show the earth while it's turning? Wouldn't that be easy to do? Yet, they haven't.

The other thing they also say is that ships disappearing over the horizon proves the earth is round, but they can't say that anymore. Cameras have advanced so much that average citizens like you and I can buy a camera that has a zoom lens powerful enough to pull that ship back into view. Get a good camera and you can do this experiment yourself.

Go to a beach and watch a ship that's far in the distance and heading for the horizon. After it disappears over the horizon, get your camera, and see if you can see that ship with your zoom lens. Voilà! That ship will amazingly come into view. It didn't drop out of sight because of the curve. People are seeing things with telescopes and zoom lenses today that they could never see before. They're seeing well beyond the fake curve.

Neville Thomas Jones, a physicist, has stated that no experiment proves that the earth is rotating. His belief is that it is stationary. He is just one of many experts that believe the same thing.

Auguste Piccard

You probably didn't hear about this guy in school, but Auguste Piccard was a Swiss physicist, inventor, and explorer. He was the first person to enter the stratosphere. He was in an airtight aluminum capsule attached to a hot-air balloon. His comment about it was "the earth seemed a flat disk with upturned edges." Fascinating.

Government Shenanigans

World War II ended in 1945. After that they started doing extensive testing and research. Exploration started in Antarctica and the decades after that involved them doing nuclear tests and shooting missiles up into space. Many believe that during the arctic explorations, they found something that freaked them out; perhaps the dome? That's why they started all their testing. If they did find a dome that kept us from leaving Earth's atmosphere, it would make sense that they would want to know the boundaries of the world we live in. How far can we go?

The powers that be won't let us go to the "South Pole." One reason is that we can't; it doesn't exist. Another reason, and this is the one I believe the most, is that they don't want us to know there is a God and the Bible is correct.

I'm going to list a time table of the testing that went on. This list will help get you started if you're wanting to do some of your own research on the governments' involvement.

Time Table of Government Testing

1945 First atomic bomb test
1946–1947 Admiral Byrd and Operation High Jump
1955–1956 Operation Deep Freeze
1957, March 11 Admiral Byrd dies[6]
1958 NASA formed
1958, February 7 DARPA The Defense Advanced Research Projects Agency was formed
1959, December 1 Antarctica Treaty System open for signature

---

[6] Ruth's Truths, https://www.bitchute.com/video/JLEzGQS9gEga/ Sending Thor to Smash the Firmament—Rob Skiba, 4-24-2022, 33:34 minutes

1961, June 23 Twelve countries sign the Antarctica Treaty System
1961 Operation Dominic (Of the Lord)[7]
1962 NASA's high altitude nuclear tests; Operation Fishbowl[8]
1970 Russia starts drilling in the Arctic Circle (The Kola Super Borehole)

Hidden in Plain Sight

It's crazy how brazen these government leaders are. They put right in front of our faces the truth and we don't see it. They're mocking us!

Look up the United Nations logo. It's a map of the flat earth!

Check out the WHO (World Health Organization) logo. Their logo is a flat earth also, but they go one step further. There's a serpent in the middle of it!

THOR (Tactical High-powered Operational Responder) actually means a god with a big hammer. Domonic means "of the Lord." (Operation Dominic and Operation Fishbowl.) They name their operations, etc. names that refer to our God and our world, then they try to deny God's existence. They think we're not smart enough to figure it out.

Truth Has Been Thrown to the Ground

A friend of mine once said, "If you're only hearing one side of the story, you're being brainwashed." This is such a true statement. It's for this very reason that a judge always has to hear the prosecutor and the defense attorney. Would he be able to give a fair sentence if only one side was allowed to speak?

---

[7] The Popular Cult https://www.bitchute.com/video/PMtX7GVCA7mQ/ The Firmament, 12-14-2023, 14:02 minutes
[8] Jerry LS, https://www.bitchute.com/video/9kuvPxY9DO8b/ Operation Dominic and Operation Fishbowl. Nuking the Flat Earth Firmament. 10-8-2022, 16:40 minutes

For the first sixty years of my life, I've only heard that the earth was round. It wasn't until 2022 that I first heard that the earth was flat. At that time, I laughed at it. It took time for me to even bother looking into it.

We can't just believe whatever we're told. We have to be discerning. We have to hear both sides. If you've only heard the earth was round all your life, you owe it to yourself to at least check the flat earth theory out and give it a reasonable amount of time and effort.

Listen to these verses God gives us about the deception that would come in the last days:

Daniel 8:12: "Because of rebellion, the LORD's people and the daily sacrifice were given over to it. It prospered in everything it did, and truth was thrown to the ground."

1 Timothy 4:1: "The Spirit clearly says that in later times some will abandon the faith and follow deceiving spirits and things taught by demons."

Why Would They Lie?

Why would they lie? This is the million-dollar question. I can't read their minds, so I can only guess at their motives.

The first possibility is all the tax-payers money they receive. In 2022, NASA received $24,000,000,000. What have they done to account for that much money? Where did all that money go? They don't have much to show for it, but that money had to go somewhere. If they get that much money each year, that would be motivation for the lie.

The second possibility is the agenda of the world leaders. Their hatred of God is apparent. Anything God says, they say the opposite. God says the earth is flat. They say it's round. God says the moon is a light. They say the moon is not a light; it's reflecting the sun's light. God says the earth is stationary and cannot be moved. They say the earth is spinning over 1000 mph and hurtling through space with the moon spinning around it, and the earth and moon are rotating around a moving sun. Wouldn't we feel all that movement? If someone has vertigo, just sitting up can send them into a spin, yet they don't feel the movement of the earth at all. Hmmm.

Christmas time was picked to celebrate the birth of Jesus, so they create a Santa Claus so we can focus on our materialistic celebrations and be distracted from the real meaning of the holiday. For Easter, they came up with the Easter Bunny and getting chocolate candy so we won't think about what Jesus did for us and how he rose from the dead. October 31st was supposed to be a celebration of the reformation of the church. They turned it into something satanic with witches, ghosts, and goblins.

Another thing the Bible says is that God created the earth and everything in it, so they say that it and everything else evolved from nothing on its own. They call this lie evolution. Their theory says there was nothing, and then magically there were two big masses. These masses one day smashed into each other and exploded. This is how the galaxies with all their stars, etc. were created. Eventually, a single cell formed and all of life grew from this one cell which appeared from nothing. All life forms and species evolved from this one cell. Now, that's a whopper of a story for you! If species evolved from another, then why don't we see lots of half cat, half dog creatures, etc.? Cats are still cats. Dogs are still dogs. I see nothing metamorphosizing into something else.

The leaders of our world have been led astray and they want a one world government. They want us to obey and follow them, not our sovereign Lord. Many of the leaders of today are satanic. Satan not only hates God, but he hates us also. When he has us thoroughly indoctrinated into his lies, we have a hard time believing the Bible when we read what it says, and that's his goal. He wants us to doubt that there is a God. He wants to be God himself. This is where I feel the real motivation lies.

The third possibility is one that I hope won't happen, but they've gone to great lengths keeping the truth from the public and feeding the lie, so I have a feeling all of this is leading to something big. I can only guess at what that is.

God's Word tells us that there's a firmament, and it's as hard as a mirror of cast bronze. The sun, moon, and all the stars are inside the firmament. There is nothing beyond the firmament but the heavens, yet they say there are many galaxies, worm holes, and who knows what else out there. I try not to spend more time than necessary listening to liars, but why has there been such an increase in talk about aliens from other planets, etc.?

Is it possible that they're going to stage in the future a fake alien attack from other planets? With all the advances that have happened at an alarming rate, such as computers and cell phones in the 1990s, why hasn't plane technology advanced that much? A passenger plane today doesn't seem much different from one from the 1960s. I think it has advanced well beyond what they're telling us. If the world sees advanced flying machines in the sky, shooting missiles and bombs, etc. at us, there will be widespread panic. We've all seen how easily the world fell into step and gave up their rights when COVID-19 was unleashed on us. If we're being attacked by "aliens," the people will do whatever they're told, even turn in their guns. They will say "We'll do anything, just save us!" It's a scary thought, but the world trusts their leaders way too much. To pull off a hoax of this magnitude you have to have the entire population convinced that there is an endless universe out there. If all of us knew we lived in a protective dome that God made for us, we would never fall for the alien lie.

A fourth possibility for the lie is that they're hiding other lands. All our world maps are made by government agencies or people they hired. International pilots have to have their flights registered and approved. There are many no-fly zones where no one is allowed, especially at the North Pole and the fake South Pole (the Antarctic ice wall).

The scary part is we don't know what they're up to or why.

What Does it Matter?

Exactly! That's what I want to know. What does it matter? Why are they spending outrageous amounts of money and effort trying to debunk flat earth? Why is it so important to them that we think we're on a ball that's spinning at an extraordinary speed? Why all the censorship? If some weirdos like me want to believe the earth is flat, why do they even care?

They've been pushing this lie at least since the 1950s when NASA was formed. Some people say that the round earth concept started 500 years ago. Before that, everyone believed it was flat. Research has been the most difficult part of writing this because it's hard knowing what's truth. I've heard different versions of when people believed the world was flat and when people started believing it was round. There are some who even claim people always believed the earth was round. The amount of deception that's out there is what's been boggling my mind the most, and this has made research of the utmost importance. When you don't know what sources you can trust, you have to be diligent.

All I know for sure is I remember hearing when I was in school that when Colombus was going to sail the oceans looking for other lands, people told him not to because he would fall off the edge of the earth. Even back then, people thought the world was flat. (I guess they didn't know about the ice wall that goes all the way around the outer edge, holding the water in.)

I'm not sure when the lie started. I'm guessing it happened over time. You can't brainwash all of Earth's inhabitants in one day. A deception of that magnitude takes time and planning.

If you're wanting to know more about when this lie started, I recommend reading *Flat Earth, Investigations into a Massive 500-Year Heliocentric Lie* by James W. Lee. He gives an amazing amount of information backing this up, and he has convinced me that this lie that the earth is a spinning ball has been around for approximately 500 years.

How Did They Fool So Many People?

It takes time and planning to fool almost everyone on the whole planet, but that's something Satan has a lot of—time.

The first step in implementing this massive deception is you have to control the information people receive. If you control everything people hear, then that's what they will believe if there are no opposing views to believe in.

Who controls the information we receive? The biggest avenues are the main stream news media, social media, our education system, politicians, world leaders, Hollywood, celebrities, the music industry, and religious leaders. The people who are wealthy and powerful have the means to join together and buy a controlling interest in the news media and corporations. Most of our news media outlets today are nothing more than propaganda machines. They are given "scripts" and told what they can and can't say. If you believe everything you hear on the news without question, then you are like a sheep being led to the slaughter. You're right where they want you.

Social media is another one that's being censored big time. I've said this once before, but I'll say it again: don't believe social media platforms that are censoring the reality of flat earth. Google and YouTube are good examples of ones to avoid. All you will get is The Flat Earth Society and people making fun of flat earth. The Flat Earth Society is a controlled opposition group that is designed to make Flat Earthers look stupid so you won't believe it or look into it. They say things like "the earth isn't spinning, it's going up!" Flat earthers don't believe that. They believe the earth is flat and stationary. It's not going up.

I wanted to mention that again because they are deceiving people. They're up to something evil or they wouldn't be putting so much effort into trying to deceive the world's population. Investigate the topic yourself and learn the truth.

The next big one is our education system. If you can teach a child from the time they're young what you want them to believe, you almost always have them for life. Our teachers and professors are good people who think they're helping the children, but they have been indoctrinated just like the rest of us. The schools are an entity of the government. People in power choose who will be deciding the curriculum that will be taught. Teachers have to follow the textbooks they're given and teach what they're told to teach. Since most of them believe in the curriculum, they have no problem doing this.

Most politicians and world leaders have no problem going along with the deception. Some have been bought and paid for by the wealthy elites, and some have been blackmailed or threatened, so they must follow the agenda. Some are wanting a one-world government, and they want to be involved in running it.

Hollywood is also owned by the wealthy and power-hungry people. Celebrities and those in the music industry have to play the game or they end up being blacklisted and no one will hire them. If you don't go along with the agenda, they can always find someone else, and you will have to give up the fame, fortune, and prestige that you've come to enjoy and relish.

The last one I'm going to cover is very sad, but true. Many of our churches and church leaders have been led astray by Satan and his deception. Folks, I can't emphasize this enough, you *must* read and study the Bible yourself to learn what is true. If reading it is a challenge for you, then listen to audio recordings of it. Start with the New Testament first because that is the most interesting and easy part of the Bible to understand. The first four books of the New Testament are the gospels, and you will learn so much about Jesus and who he is. You will be amazed.

Satan is clever and he attacks the churches and purposely tries to deceive them and lead them away from God. The God of the Bible is a Triune God. The Father, Son, and Holy Spirit are separate, yet they are one. This is hard for us to understand, but it's what the Bible tells us.

Unfortunately, some churches teach that Mary is God. This is one of Satan's biggest deceptions. Run from these teachings! The bible says that Mary was afraid of the angel that appeared to her. A God would not be afraid of an angel. The bible also says that after the miraculous virgin birth of Jesus, Mary had other children with Joseph. Would a God marry a human and have children with him? No, they would not. Do not worship anything or anyone besides the one true God who is the Father, Son, and Holy Spirit.

I know in the church I go to, the pastors love God and teach from the Bible, but even they have been led astray on certain things. Yes, they worship the one true God and him only, which is good, but they've been misled just like I was. They believe we live on a spinning ball hurtling through space. While believing this lie won't keep you out of heaven, it's still a harmful lie. If you believe we live on a spinning ball and the Bible tells you we live on a flat, stationary earth, who will you believe? Unfortunately, many will believe the lie and they'll turn from God. They'll say the Bible is nonsense. At best, they will say things like "Well, that's not what the Bible meant to say. It isn't literal." When you start doubting the Bible, your faith is on shaky ground, and that's just where the Devil wants you! God means every word he says. The Bible is a gift from God to us so that we can gain wisdom and learn about our Lord and Savior, Jesus Christ. Believe in God and follow him and you will be assured a permanent home with him in heaven. I'm looking forward to that!

God's Perfect Creation

I'm guessing this will be my favorite part of this book to write. God's creation. Wow! Let's just start with us little humans.

I just finished reading a book called *Creating a Better Brain through Neuroplasticity* by Debi Pearl. Just our brain alone is beyond all understanding. If part of the brain gets damaged, it can reroute and send the information needed through another path. If one of our senses is lost or damaged, our other senses increase in their abilities to compensate for this loss. I also looked up information on the internet. Our brain's storage capacity is virtually unlimited. Brain information travels up to 268 miles per hour! The human brain contains approximately one hundred billion neurons. Amazing!

I just touched on a few things about the human brain, and we have the rest of our body to think about too: our heart, digestive system, immune system, etc. We are fearfully and wonderfully made.

Humans are just one of God's many creations. Think of all the different species of plants, birds, animals, insects, etc. Everything in nature is so extraordinary in how it works and functions that it boggles the brain (but your brain can handle it, because God made it to be an amazing organ)!

Now we get to our earth. It is flat. (If you don't agree, humor me here please.) The North Star, Polaris, is in the middle, above the North Pole. It always stays put. It never moves from that location. The other stars circle around it. There are many constellations. Many of us are familiar with the twelve we always hear about: Aries, Taurus, Gemini, Cancer, Leo, Virgo, Libra, Scorpio, Sagittarius, Capricorn, Aquarius, Pisces. The constellation we see in January, we will see next January too. Everything is predictable and precise. Everything is not randomly hurtling through space, sometimes smashing into each other. We're also not spinning. God would never put us on a ball that's spinning. What would be the purpose in that? Everything about God's creation makes perfect sense.

Another thing is eclipses. When they say that we are going to have an eclipse because the moon will pass in front of the sun, you get a "ring of fire." This ring is perfect because the sun and moon are near each other and they are the exact same size and shape. It's hard to image that a big bang created such perfection.

When God made this world, it was perfect, and it was good. Mankind has made a mess of things, but there is still much of the beauty and miraculous wonders that were here in the beginning. It boggles my mind when I try to imagine what heaven will be like, because it's supposed to be many times more wonderful than the most beautiful place here on Earth. I can't even imagine what's in store for us.

Romans 1:20: "For since the creation of the world God's invisible qualities—his eternal power and divine nature—have been clearly seen, being understood from what has been made, so that people are without excuse."

Researching Flat Earth for Yourself

If you're still convinced that the earth is round, I hope you will at least spend some time researching flat earth. (Since you're reading this, I have hope that you will.)

You've probably heard the earth is round all your life. To come to a logical conclusion, you must give both sides a listen.

I'll give you some pointers and places to go, but first I'll tell you where not to go. Don't do any searches on YouTube, Google, or Facebook. They're censoring this subject big time, which amazes me. Why would they even care if a few crazies like me believe the earth is flat? Why go to these lengths to debunk it?

Another thing they did is create a fake flat earth group called The Flat Earth Society. This was set up so they could have them say ridiculous stuff and make true flat-earthers look stupid. One thing they say is that the earth isn't spinning, it's going up! No true flat earther believes that. The earth is stationary. It does not move.

A couple platforms that I know of that don't censor it are Rumble.com and Bitchute.com. I'm sure there are more, and the internet is constantly changing. On Bitchute.com you will type in flat earth in the search bar. It automatically comes up under "relevance." You can change the setting from relevance to newest and get many other options of videos to choose from. (However, "200 proofs the earth is not a spinning ball" is a good one to watch under relevance).

These are some of the people and sites that have information on the internet: Rob Skiba (now deceased, but his stuff is still out there), Conspiracy Music Guru (for fun music), Mark Sargent, Dean Odle, Celebrate Truth, and Robbie Davidson.

A good reason to research flat earth:

John 8:32: "Then you will know the truth, and the truth will set you free."

My Journey from Round to Flat

Like most people, I always believed what I was taught at school and what I heard from our government and on TV. (We can't blame teachers. They were also deceived. They thought they were teaching the truth.)

One thing I never believed in was evolution, but I believed the earth was round and spinning even though I couldn't see it or feel it.

The earth was flat? I never even heard it suggested until 2022. My indoctrination of their round earth was instilled so deep that I never even noticed all the verses in the Bible that support it being flat! Over the last forty plus years, I've read the Bible numerous times. I would zip right on by all those verses, not absorbing what it was saying. That's how deep my indoctrination had been. Now I'm just flabbergasted when I come across those verses and wonder how I missed them before.

In January of 2023, the subject of flat earth was brought up to me once again. This time I decided to look into it with an open mind. Wow! It was mind blowing. I couldn't believe it.

What's funny is the thought that the sun and moon aren't that far away and that God is above them has brought me such peace and joy. When I look at the moon at night, I get a big smile on my face. I'll look up at it and think, "You're not that far away, are you?" Then the man on the moon will smile and wink at me. Okay, that last part isn't true, but that would be neat if it were!

This discovery has opened my eyes to everything. I used to get scientific discoveries sent to my email. They would talk about worm holes, the poles flipping, and all kinds of weird things, and I would just believe it! They would also show a rock or something they found and say something like "this is 847,245,327 years old." How do they know that? Ridiculous. I now just laugh when I hear stuff like that, so I eventually canceled those subscriptions.

The deception in this world has gotten so bad that I pretty much don't believe anything I hear anymore, unless there's real proof and it gels with what the Bible says. Yahweh is my LORD, and I believe in him with all my heart. His Word has never let me down.

My prayer for all of you is that your journey is as amazing as mine was and that you find yourself feeling even closer to God than you did before.

God bless you!

RESOURCES AND REFERENCES

These are just some of the books I read and videos I watched while doing my research. Some things we will never know for sure because they won't let anyone go to the poles to see for themselves. I also can't guarantee these resources will still be out there and available by the time you read this. Again, don't research on censored platforms (Google, YouTube, etc.) You'll just get more brainwashing.

Books

Flat Earth: Investigations into a Massive 500-year Heliocentric Lie by James W. Lee
Copyright 2017, ISBN-10:1542805333, ISBN: 978-1542805339
Flat Earth: How Science Replaced God and Deceived the Whole World by Caspian Sarginson, J.D. Copyright 2023, ISBN: 9798397796156
Flat Earth and Satan's Global Plan of Deception: Heliocentrism Exposed by Caspian Sarginson Copyright 2023, ISBN: 9798374256772

Videos

Celebrate Truth. https://rumble.com/v3is8vs--nasa-lies-and-other-space-deceptions-flat-earth-247.html. "NASA lies and Other Space Deceptions"/Flat Earth 24/7 September 18th, 25:00:07 minutes
Dubay, Eric. https://www.bitchute.com/video/XvAwLc7FZm2z/. "200 Proofs earth is not a spinning ball" August 18, 2018 1:29:03 minutes
Flat Earth Films. https://rumble.com/vg3733-ex-nasa-employee-matt-boylan-aka-math-powerland-discusses-flat-earth.html. "Ex NASA

employee Matt Boylan aka Math Powerland discusses Flat Earth." 2022, 14:02 minutes

Phiroc. https://www.bitchute.com/video/uCQhOxLsmLLE/. "The 35 most common flat earth questions answered in 35 minutes." 2-5-24, 35:57 minutes

Rico Rozy (aka Black D). https://www.bitchute.com/video/F0Eh 1rX7NQ99/. "Only on a Flat Earth Water can find His Level." 1-13-2024, 1:10:11 minutes

Rose Arcana_2023. https://www.bitchute.com/video/qWjZ8gjRmY-Nx/. "Flat Earth/All Meat & no Potatoes." 1-17-2024, 57:16 minutes

Truth Conspiracy Reality. https://www.bitchute.com/video/QmOP1QOpbdIE/. "We Don't have the Technology to do that anymore, we used to, but we destroyed it." 11-15-2020, 0:14 minutes

Truth Conspiracy Reality. https://www.bitchute.com/video/Q7asRm HLDC12/. "NASA's Robert Simmon Admits Ball Earth Images are Photoshopped! The Earth is Flat." 11-15-2020, 1:53 minutes

WarriorForChrist777https://www.tiktok.com/@ephesians6_10/video/7139 407564227366187?lang=en

Ephesians 6_10, 9-4-2022, 3:04 Minutes (1958 encyclopedia Britannica. Antarctica.)

Wolf Watch: Impossiball, https://totaldisclosure.net/impossiball/. "Total Disclosure Films." 2017, 1:34:27 minutes

ABOUT THE AUTHOR

Mary K. Eastman has always dreamed of being an author. In her latter years, she decided to stop dreaming and start doing. In 2019 her suspense romance novel, *Return to Sleeping Bear,* came out. In 2020, during the pandemic, her interests turned to seeking out truth and finding out what was really going on in the world. Feeling an overwhelming need to share what she had learned, in spite of the opposition she knew it would bring her, she forged ahead and finished *The Truth About our World.*

Mary has been a Michigander all her life. Being a bit of a water fanatic, she feels like a fish in water living in Michigan and never wants to live anywhere else.

The things most important to her are her faith in God, her family, and her friends. She loves being a mom and grandma. Some things she enjoys doing are being in the outdoors, reading, writing, sailing, white-water rafting, kayaking, reading God's Word, and spending time with those whom she loves.

You can find her on the following:

Blog: michiganmary.com

Website: marykeastman.com